HEMP

The Plant That Will Help Save the World

In a world grappling with climate change, deforestation, pollution, and dwindling natural resources, one plant stands out as a beacon of hope: hemp. Often misunderstood and historically stigmatized, hemp is a versatile and sustainable crop that has the potential to address many of the pressing challenges we face today. This book explores the myriad ways in which hemp can contribute to a more sustainable and equitable future.

Table of Contents

Chapter 1: A Brief History of Hemp 9
- Ancient Beginnings 9
- Spread Across the Globe 9
- The New World 10
- The Fall from Grace 10
- The Resurgence 11

Chapter 2: Hemp and the Environment 14
2.1 Carbon Sequestration 14
- Understanding Carbon Sequestration 14
- Lifecycle of Carbon in Hemp Cultivation 15
- The Role of Hemp in Sustainable Agriculture 16
- Real-World Impact 17
- Challenges and Considerations 17

2.2 Soil Health 20
- Enhancing Soil Fertility 20
- Reducing Soil Erosion 21
- Soil Microbial Life 22
- Sustainable Farming Practices 23
- Case Studies and Success Stories 24

2.3 Water Conservation 26
- Low Water Requirements 26

Drought Resistance 27
Soil Moisture Retention 27
Reducing Irrigation Needs 28
Challenges and Considerations 29
Chapter 3: Hemp as a Renewable Resource 31
3.1 Paper Production 31
The Environmental Impact of Traditional Paper Production . 31
The Advantages of Hemp Paper 32
Economic Benefits 34
Case Studies and Success Stories 34
Challenges and Considerations 35
3.2 Building Materials 37
What is Hempcrete? 37
Environmental Benefits 38
Performance Characteristics 39
Case Studies and Real-World Applications 41
Challenges and Considerations 41
3.3 Bioplastics 44
Understanding Bioplastics 44
The Advantages of Hemp-Based Bioplastics 45
Types of Hemp-Based Bioplastics 46
Real-World Applications 47

    Challenges and Considerations 48

Chapter 4: Hemp in Agriculture 50

  4.1 Crop Rotation...50

    The Importance of Crop Rotation 50

    Integrating Hemp into Crop Rotation 52

    Case Studies and Success Stories..53

    Challenges and Considerations 54

  4.2 Animal Feed...56

    Nutritional Profile of Hemp Seeds..56

    Benefits of Hemp in Livestock Farming.................................57

    Applications in Different Livestock..59

    Challenges and Considerations 60

Chapter 5: Hemp for Human Health..62

  5.1 Nutritional Benefits..62

    Nutritional Profile of Hemp Seeds..62

    Health Benefits of Incorporating Hemp into Your Diet.........64

    Ways to Incorporate Hemp into Your Diet 66

  5.2 Medicinal Uses 68

    Understanding CBD and Its Properties 68

    Potential Health Benefits of Hemp-Derived CBD 69

    Methods of Consumption..71

    Challenges and Considerations 72

Chapter 6: Economic Opportunities 75
6.1 Job Creation 75
The Expanding Hemp Market 75
Job Opportunities in the Hemp Industry 76
Economic Impact on Local Communities 78
Case Studies and Success Stories 79
Challenges and Considerations 80
6.2 Global Trade 83
The Growing Global Market for Hemp 83
Key Hemp Products in Global Trade 85
Opportunities for Economic Diversification 86
Challenges and Considerations 88
Chapter 7: Policy and Regulation 90
7.1 Legal Challenges 90
Historical Context 90
Current Legal Challenges 91
The Need for Policy Reform 93
Case Studies of Successful Legal Reforms 95
7.2 Advocacy and Education 97
The Importance of Advocacy 97
Educational Initiatives 99
Successful Advocacy Examples 100

Challenges to Advocacy 102
Increasing Awareness Through Collective Action 103
One Effective Strategy: Sharing Knowledge Through This Book 104
Expected Impact of Sharing This Book 105

Hemp is a practical solution 107
--- References 109
Books 109
Articles and Journals 110
Reports and White Papers 111
Websites and Online Resources 112

Appendix 114
Recipes Using Hemp 114
DIY Projects with Hemp 116
Hemp-Related Organizations and Businesses 117
Organizations 117
Businesses 118

The Plant That Will Help Save the World 121
1. Elevating Awareness 123
2. Personal Advocacy 123
3. Encouraging Policy Reform 124
4. Building Momentum 124

5. Demonstrating Public Support 125

6. Fostering Collaboration..125

7. Sending a letter to your government officials 125

Here is a sample letter you can use or create your own. 126

Chapter 1: A Brief History of Hemp

Ancient Beginnings

Hemp, known scientifically as Cannabis sativa, is one of the oldest cultivated crops in human history. Archaeological evidence suggests that hemp was first used around 8,000 BCE in ancient China, where it was cultivated for its strong fibers and nutritious seeds. The ancient Chinese used hemp to make ropes, textiles, and paper. Hemp was so integral to Chinese society that it was even mentioned in ancient texts, including the earliest known pharmacopoeia, the Shennong Bencao Jing.

Spread Across the Globe

From China, hemp cultivation spread to other parts of Asia, the Middle East, and eventually to Europe. In India, hemp was used in traditional Ayurvedic medicine and as a source of fiber for clothing and ropes. The ancient Egyptians also utilized hemp for various purposes, including making sails for their boats.

By the time hemp reached Europe, it had become a staple crop. The ancient Greeks and Romans used hemp for a variety of applications, including ropes, sails, and clothing. The Romans even used hemp seeds as a food source. Throughout the Middle Ages, hemp continued to be a crucial crop in Europe, providing essential materials for daily life.

The New World

When European settlers arrived in the Americas, they brought hemp seeds with them. Hemp quickly became an important crop in the New World, particularly in colonial America. The early American colonies mandated hemp cultivation, recognizing its value for producing ropes, sails, and clothing. In fact, both George Washington and Thomas Jefferson were known to have grown hemp on their estates.

The Fall from Grace

Despite its long history of utility, hemp's reputation began to suffer in the 20th century. The primary reason for this fall from grace was its association with marijuana, a psychoactive variant of the Cannabis plant. In the 1930s, the United States launched a campaign against marijuana, driven by racial and political factors. Unfortunately, hemp, despite its non-psychoactive properties, was caught in the crossfire.

The 1937 Marihuana Tax Act effectively criminalized all forms of Cannabis, including hemp. This legislation, coupled with the rise of synthetic fibers and other industrial materials, led to a sharp decline in hemp cultivation. During World War II, there was a brief resurgence in hemp production in the United States through the "Hemp for Victory" campaign, which encouraged farmers to grow hemp for the war effort. However, this was short-lived, and hemp production dwindled again after the war.

The Resurgence

In recent decades, there has been a renewed interest in hemp, driven by growing awareness of its environmental and economic benefits. The 2018 Farm Bill in the United States marked a significant turning point by legalizing the cultivation of hemp with less than 0.3% THC, the psychoactive compound found in marijuana. This legislation opened the door for farmers to once again grow hemp and for researchers to explore its myriad uses.

Today, hemp is experiencing a renaissance. Advances in technology have made it possible to process hemp more efficiently, unlocking new applications in industries ranging from textiles and construction to food and medicine. Countries around the world are revisiting their policies on hemp, recognizing its potential to contribute to a more sustainable and prosperous future.

Hemp's journey from ancient crop to modern wonder plant is a testament to its enduring value and versatility. Despite the setbacks it faced in the 20th century, hemp is making a triumphant return, offering solutions to some of the most pressing challenges of our time. As we move forward, understanding the history of hemp can help us appreciate its potential and guide us in harnessing its benefits for a better world.

In the chapters that follow, we will delve deeper into the various ways hemp can help address environmental, economic, and social challenges, reaffirming its place as a plant that can help save the world.

Chapter 2: Hemp and the Environment

2.1 Carbon Sequestration

As the world grapples with the devastating impacts of climate change, the need for effective strategies to mitigate greenhouse gas emissions has never been more urgent. Among these strategies, carbon sequestration—the process of capturing and storing atmospheric carbon dioxide (CO_2)—plays a crucial role. Hemp, a fast-growing and versatile plant, has emerged as a potent tool in this endeavor.

Understanding Carbon Sequestration

Carbon sequestration occurs naturally through various processes, including photosynthesis, where plants absorb CO_2 from the atmosphere and convert it into biomass—leaves, stems, roots, and flowers. This process not only helps reduce the concentration of greenhouse gases but also supports the growth and health of the plant.

Hemp is particularly effective in this regard due to its rapid growth rate and high biomass yield. Studies indicate that hemp can absorb up to 22 tons of CO_2 per hectare (approximately 2.47 acres) over a growing season, making it one of the most efficient crops for carbon capture.

Lifecycle of Carbon in Hemp Cultivation

When hemp is cultivated, it captures CO_2 during its growth phase. The carbon absorbed by the plant is stored in its biomass. Once harvested, the carbon can remain stored in various forms depending on how the hemp is utilized:

1. Biomass Utilization: If hemp is processed into products like paper, textiles, or biocomposites, the carbon remains sequestered in these materials for extended periods. For example, hemp fibers used in construction materials can sequester carbon for decades.

2. Soil Carbon Storage: The roots of hemp enhance soil health and structure, which can lead to increased soil organic carbon levels. When hemp is grown, its root systems help improve soil aeration and water retention, fostering a thriving ecosystem in the soil that can store additional carbon.

3. Biomass Conversion: Hemp can also be converted into biofuels or biochar. Biochar is a carbon-rich product created by pyrolyzing biomass in a low-oxygen environment, which can sequester carbon in the soil for hundreds to thousands of years.

The Role of Hemp in Sustainable Agriculture

Hemp's ability to sequester carbon is amplified when integrated into sustainable agricultural practices. Crop rotation, where hemp is alternated with other crops, can enhance soil fertility and health while improving carbon storage. By reducing the need for chemical fertilizers and pesticides, hemp cultivation can also minimize soil degradation and pollutant runoff, further benefitting the environment.

Real-World Impact

Several studies and pilot projects have demonstrated the potential of hemp as a carbon sink. For example, researchers in Canada have documented significant carbon sequestration in hemp fields compared to traditional crops. Additionally, countries like France and the Netherlands have begun to promote hemp cultivation as a strategy to meet their climate goals.

By incorporating hemp into agricultural policies and practices, governments can capitalize on its benefits for carbon sequestration while supporting farmers and rural communities.

Challenges and Considerations

While the potential for hemp in carbon sequestration is promising, it is important to consider the broader context of land use and agricultural practices. Sustainable farming methods must be employed to ensure that hemp cultivation does not lead to adverse environmental impacts, such as biodiversity loss or soil depletion.

Furthermore, the promotion of hemp should be part of a comprehensive approach to climate change that includes reducing fossil fuel use, enhancing energy efficiency, and protecting natural ecosystems.

Hemp stands out as a powerful tool in the fight against climate change, offering a natural solution for carbon sequestration. Its rapid growth, high biomass yield, and versatility make it an invaluable crop for sustainable agriculture. As we continue to explore the environmental benefits of hemp, it becomes increasingly clear that this ancient plant has a crucial role to play in creating a sustainable future for our planet.

In the following sections, we will delve into other environmental benefits of hemp, including its potential for improving soil health, conserving water, and providing alternative materials to reduce reliance on fossil fuels and timber. Together, these attributes position hemp not just as a crop of the past, but as a cornerstone of a more sustainable and resilient future.

2.2 Soil Health

The health of our soil is a foundational element of sustainable agriculture and environmental stewardship. Healthy soil supports plant growth, regulates water cycles, and serves as a habitat for diverse organisms. However, modern agricultural practices have led to soil degradation, erosion, and the overuse of chemical fertilizers, which can harm both the environment and human health. Hemp, with its unique properties and cultivation practices, offers a pathway to enhance soil health while promoting sustainable farming.

Enhancing Soil Fertility

One of the most significant benefits of growing hemp is its ability to enrich soil fertility. Hemp has a deep and extensive root system that penetrates the soil, aerating it and improving its structure. This natural process enhances water infiltration and reduces surface runoff, allowing more water to reach the root zone of plants.

Moreover, hemp is a dynamic accumulator, meaning it has the ability to absorb and concentrate nutrients from the soil. As it grows, hemp draws up essential minerals, including nitrogen, phosphorus, and potassium, which are vital for plant health. When hemp is harvested, its biomass returns these nutrients to the soil, enriching it for the next crop. This cyclical process reduces the need for synthetic fertilizers and helps maintain soil fertility over time.

Reducing Soil Erosion

Soil erosion is a significant environmental concern, leading to the loss of arable land and the degradation of ecosystems. Traditional agricultural practices, such as monocropping and excessive tillage, can exacerbate erosion by disturbing the soil structure and removing protective vegetation.

Hemp, with its robust root system, helps to anchor the soil in place, preventing erosion caused by wind and water. The plant's dense foliage also provides ground cover, shielding the soil from the impact of heavy rainfall and reducing the risk of erosion. By incorporating hemp into crop rotation and cover cropping systems, farmers can effectively combat soil erosion and preserve valuable topsoil.

Soil Microbial Life

Healthy soil is teeming with life, including beneficial microorganisms that contribute to nutrient cycling and plant health. Hemp cultivation fosters a diverse soil ecosystem by promoting microbial activity. The plant's root exudates—organic compounds secreted by roots—serve as food for soil microbes, enhancing their populations and activity.

This increase in microbial diversity can lead to improved soil health and fertility, as these microorganisms play critical roles in breaking down organic matter and making nutrients available to plants. Healthy soil ecosystems also enhance resilience to pests and diseases, reducing the need for chemical pesticides.

Sustainable Farming Practices

Integrating hemp into sustainable farming practices can have far-reaching benefits for soil health. Here are some strategies that farmers can adopt:

1. Crop Rotation: By rotating hemp with other crops, farmers can enhance soil fertility and reduce the buildup of pests and diseases. Hemp's ability to draw nutrients from deep in the soil also complements shallower-rooted crops.

2. Cover Cropping: Planting hemp as a cover crop during the off-season helps protect the soil from erosion while improving its structure and nutrient content. Cover crops also suppress weeds and enhance biodiversity.

3. Minimal Tillage: Reducing tillage minimizes soil disturbance, preserving soil structure and microbial life. Hemp's deep roots can help maintain soil aeration and prevent compaction.

4. Organic Fertilization: Farmers can rely on organic amendments, such as compost or manure, to enhance soil

fertility in conjunction with hemp cultivation. This practice reduces reliance on synthetic fertilizers and promotes a healthier soil ecosystem.

Case Studies and Success Stories

Several farmers and agricultural initiatives worldwide have successfully integrated hemp into their practices, demonstrating its benefits for soil health. For example, in Canada, hemp has been used in crop rotation systems to restore soil fertility in fields previously depleted by conventional farming methods. Farmers report improved yields and healthier soil after incorporating hemp into their rotations.

Similarly, in Europe, organic farmers have embraced hemp as a cover crop to improve soil quality and reduce the need for chemical inputs. These practices have shown promising results in enhancing soil structure and biodiversity, contributing to more sustainable agricultural systems.

Hemp is not just a crop; it is a powerful ally in the quest for sustainable farming practices and soil health. By enriching the soil, reducing erosion, and fostering a diverse microbial ecosystem, hemp cultivation can significantly enhance agricultural sustainability. As farmers and policymakers recognize the importance of healthy soils for food security and environmental resilience, hemp stands out as a viable solution for promoting sustainable agriculture.

In the subsequent sections, we will explore more environmental benefits of hemp, including its role in water conservation and its potential to provide alternative materials that can reduce our reliance on fossil fuels and timber. Together, these attributes solidify hemp's position as a crucial player in nurturing a more sustainable planet.

2.3 Water Conservation

Water is one of our planet's most precious and limited resources. As global demand for freshwater continues to rise due to population growth, agricultural expansion, and climate change, the need for sustainable water management practices has never been more critical. Hemp, a resilient and adaptable crop, has emerged as a key player in promoting water conservation in agricultural systems.

Low Water Requirements

One of the most remarkable attributes of hemp is its relatively low water requirement compared to many traditional crops. While the exact water needs can vary based on factors such as soil type, climate, and growth stage, hemp generally requires about half the amount of water that crops like cotton or corn need. This efficiency makes hemp an attractive option for farmers in regions facing water scarcity or those looking to adopt more sustainable agricultural practices.

Drought Resistance

Hemp is known for its resilience to drought conditions. Its deep root system allows it to access moisture from deeper soil layers, enabling the plant to thrive even when surface water is limited. This ability not only supports the growth of hemp but also helps preserve the moisture levels in the topsoil, benefiting subsequent crops in rotation.

By cultivating hemp, farmers can maintain productivity while reducing their overall water usage. This is especially important in arid and semi-arid regions, where water resources are increasingly strained.

Soil Moisture Retention

Hemp contributes to water conservation not only through its low water requirements but also by enhancing soil moisture retention. The dense canopy of hemp plants provides shade to the soil, reducing evaporation rates during hot weather. Additionally, the extensive root system improves soil structure, promoting better water infiltration and retention.

When hemp is grown as a cover crop or in rotation with other crops, it can help maintain soil moisture levels, leading to healthier and more productive growing conditions. This is particularly beneficial in areas prone to drought, as it allows farmers to maximize their yields while minimizing water consumption.

Reducing Irrigation Needs

The cultivation of hemp can also lead to reduced irrigation needs for surrounding crops. By integrating hemp into agricultural systems, farmers can create a more diverse cropping environment that benefits overall water management. For example, when hemp is used as a cover crop, it can improve soil health and moisture retention, which in turn supports the growth of other crops.

Moreover, hemp's ability to suppress weeds and provide ground cover can reduce the competition for water among plants, allowing for more efficient water use. This practice can lead to significant water savings over time, particularly in conjunction with other sustainable farming practices.

Case Studies and Real-World Applications

Farmers around the world are beginning to recognize the water-saving potential of hemp. In the United States, researchers have conducted trials in arid regions to assess hemp's water efficiency compared to traditional crops. The results have shown that hemp can maintain competitive yields while using significantly less water, making it an ideal candidate for sustainable agriculture in water-limited environments.

In Europe, farmers are adopting hemp in crop rotation systems to optimize water use. By interspersing hemp with other crops, they have reported improved soil moisture retention and reduced irrigation needs, leading to more resilient farming systems.

Challenges and Considerations

While hemp offers numerous benefits for water conservation, it is essential to consider the broader context of agricultural practices and water management. Effective water conservation requires a holistic approach that includes soil health, crop diversity, and efficient irrigation techniques.

Additionally, local climate conditions and water availability should be taken into account when deciding on hemp cultivation. Farmers must also be educated on best practices for growing hemp sustainably to maximize its benefits for water conservation.

Hemp stands out as a champion for water conservation in agricultural systems. Its low water requirements, drought resistance, and ability to enhance soil moisture retention make it an invaluable crop for sustainable farming practices. As water scarcity becomes an increasingly pressing issue worldwide, the cultivation of hemp can contribute significantly to more efficient water use and management.

In the following sections, we will continue to explore the environmental benefits of hemp, including its role in reducing reliance on synthetic materials and fossil fuels, further solidifying its position as a sustainable resource for a healthier planet.

Chapter 3: Hemp as a Renewable Resource

3.1 Paper Production

The production of paper has long been associated with deforestation and environmental degradation. Traditional paper manufacturing relies heavily on wood pulp, leading to the destruction of forests, loss of biodiversity, and increased carbon emissions. As the demand for paper continues to grow in an increasingly paper-dependent world, the need for sustainable alternatives has become more critical than ever. Hemp, with its unique properties and cultivation practices, offers a compelling solution.

The Environmental Impact of Traditional Paper Production

The traditional paper industry is responsible for significant deforestation, particularly in ecologically sensitive regions. The harvesting of trees for paper production contributes to habitat loss, soil erosion, and disruption of local ecosystems. Additionally, the chemical processes involved in turning wood into paper can lead to water pollution and contribute to climate change through the release of greenhouse gases.

According to the World Wildlife Fund, approximately 40% of the world's industrial wood harvest is used for paper production. This high demand places immense pressure on forests, resulting in the depletion of natural resources and threatening the planet's biodiversity.

The Advantages of Hemp Paper

Hemp paper presents a sustainable alternative to traditional wood pulp paper. Here are some of the key benefits of switching to hemp-based paper:

1. Sustainability: Hemp is a fast-growing, renewable resource that can produce multiple harvests per year. In contrast to trees, which can take decades to mature, hemp reaches full height in just 3 to 4 months. This rapid growth cycle allows for sustainable harvesting without depleting resources.

2. Less Land and Water Use: Hemp requires significantly less land and water compared to tree farming for paper. Studies have shown that hemp can yield 4 to 10 times more fiber per acre

than trees, making it a more efficient crop for paper production. Additionally, its lower water requirements contribute to better resource management.

3. Durability and Quality: Hemp fibers are longer and stronger than those found in wood pulp, resulting in a more durable and high-quality paper product. Hemp paper is less likely to yellow or degrade over time, making it ideal for archival purposes and long-lasting documents.

4. Reduced Chemical Use: The process of making hemp paper typically requires fewer chemicals than traditional wood pulp paper production. Hemp fibers can be processed using more environmentally friendly methods, leading to less toxic waste and reduced water pollution.

5. Carbon Sequestration: As previously discussed, hemp cultivation contributes to carbon sequestration. By switching to hemp paper, we not only reduce our reliance on wood but also support the growth of a crop that actively absorbs CO_2 from the atmosphere.

Economic Benefits

In addition to its environmental advantages, hemp paper production can also create economic opportunities. As demand for sustainable products rises, investing in hemp paper manufacturing can stimulate local economies and create jobs in farming, processing, and distribution.

The hemp industry is also becoming increasingly recognized for its potential to drive innovation in various sectors, including packaging, textiles, and construction. By establishing a hemp paper market, businesses can diversify their product offerings and attract environmentally conscious consumers.

Case Studies and Success Stories

Several companies and organizations are already making strides in the hemp paper industry. For instance, in the United States, there are initiatives to revive hemp cultivation and promote hemp paper production. Some businesses have successfully developed hemp-based products, including stationery, packaging, and even banknotes.

In Canada, hemp paper manufacturers have emerged, producing high-quality paper products while promoting sustainable agricultural practices. These efforts demonstrate the viability of hemp as a sustainable alternative to traditional paper and showcase the growing interest in environmentally friendly materials.

Challenges and Considerations

While the transition to hemp paper presents numerous benefits, there are challenges to consider. One major hurdle is the existing infrastructure for paper production, which is heavily geared toward wood pulp. Developing new processing facilities and supply chains for hemp paper will require investment and collaboration among stakeholders.

Additionally, educating consumers and businesses about the benefits of hemp paper is essential to drive demand and acceptance. As awareness grows, more people may choose hemp-based products over traditional paper, paving the way for a more sustainable paper industry.

Hemp paper offers a sustainable and environmentally friendly alternative to traditional wood pulp paper. By switching to hemp-based products, we can help reduce deforestation, conserve natural resources, and promote a healthier planet. The durability and quality of hemp paper, combined with its lower environmental impact, make it a compelling choice for consumers and businesses alike.

As we continue to explore the renewable potential of hemp, the next sections will delve into other applications of this versatile plant, including its use in building materials and bioplastics—further emphasizing hemp's role in creating a sustainable future.

3.2 Building Materials

As the construction industry faces increasing scrutiny over its environmental impact, there is an urgent need for sustainable building practices that minimize resource consumption and reduce carbon emissions. Among the most promising solutions is hempcrete—a composite material made from hemp hurds (the inner woody core of the hemp plant) and lime. This innovative building material is not only highly sustainable but also offers exceptional performance characteristics that can revolutionize the construction industry.

What is Hempcrete?

Hempcrete is a biocomposite material that combines hemp hurds with lime-based binders. The resulting material is lightweight, highly insulating, and capable of regulating humidity. Unlike traditional concrete, which is heavy and energy-intensive to produce, hempcrete is made from renewable resources and has a much lower environmental footprint.

Hempcrete is typically used as infill in timber-framed structures, providing insulation and thermal mass without bearing structural loads. This unique application allows for energy-efficient building designs that can significantly reduce heating and cooling costs.

Environmental Benefits

1. Sustainable Sourcing: Hemp is a fast-growing crop that can be harvested multiple times a year. It requires minimal pesticides and fertilizers, making it a sustainable choice for building materials. The use of hemp in construction helps promote sustainable agriculture and land use.

2. Carbon Sequestration: As discussed in previous chapters, hemp plants absorb carbon dioxide during their growth. When used in building materials like hempcrete, the carbon remains sequestered, contributing to carbon-neutral or even carbon-negative construction practices.

3. Reduced Energy Consumption: The production of traditional building materials, such as concrete and steel, is highly energy-intensive and contributes significantly to greenhouse gas emissions. Hempcrete, on the other hand, requires considerably less energy to produce. Additionally, the excellent insulation properties of hempcrete help regulate indoor temperatures,

reducing the reliance on heating and cooling systems and further lowering energy consumption.

4. Biodegradability: Hempcrete is made from natural materials, making it biodegradable at the end of its life cycle. Unlike conventional building materials that can contribute to landfill waste, hempcrete can return to the earth without causing environmental harm.

Performance Characteristics

1. Insulation: Hempcrete is an excellent insulator, providing superior thermal performance compared to traditional materials. Its insulating properties help maintain comfortable indoor temperatures, reduce energy costs, and minimize reliance on fossil fuels for heating and cooling.

2. Humidity Regulation: One of the unique features of hempcrete is its ability to regulate humidity levels within buildings. The porous structure of the material allows it to absorb excess moisture from the air, helping to maintain a balanced indoor environment. This property can reduce the risk of mold growth and enhance indoor air quality.

3. Fire Resistance: Hempcrete is naturally fire-resistant due to its lime content. This characteristic enhances the safety of buildings constructed with hempcrete, making it an attractive option for homeowners and builders alike.

4. Acoustic Performance: The dense and fibrous nature of hempcrete provides good sound insulation, making it an ideal choice for residential and commercial buildings where noise reduction is a priority.

Case Studies and Real-World Applications

The use of hempcrete is gaining traction in various parts of the world. In Europe, countries like France and the Netherlands have embraced hempcrete as a sustainable building material. Numerous projects, including residential homes, schools, and community centers, have successfully utilized hempcrete for its environmental benefits and performance characteristics.

In the United States, interest in hempcrete is growing, with architects and builders exploring its potential in both new construction and retrofitting existing structures. Several organizations are conducting research and development to optimize hempcrete formulations and promote its use in modern building practices.

Challenges and Considerations

Despite its many advantages, the widespread adoption of hempcrete faces several challenges. One major hurdle is the need for standardized building codes and regulations that recognize hempcrete as a legitimate construction material. As with any building product, ensuring compliance with safety and performance standards is essential for gaining acceptance in the construction industry.

Additionally, there may be a learning curve for builders and contractors unfamiliar with hempcrete construction techniques. Education and training programs will be crucial to equip industry professionals with the knowledge and skills needed to work with this innovative material effectively.

Hempcrete represents a transformative opportunity for the construction industry, offering a sustainable, high-performance alternative to traditional building materials. By harnessing the environmental benefits of hemp, we can create energy-efficient, durable, and aesthetically pleasing structures that contribute to a more sustainable future.

As we continue to explore the renewable potential of hemp, the next sections will delve into other applications of this versatile plant, including its use in bioplastics and other innovative products that can help reduce our reliance on fossil fuels and enhance sustainability across various industries.

3.3 Bioplastics

In recent years, the environmental crisis posed by plastic pollution has reached alarming levels. Single-use plastics, which dominate packaging and consumer products, contribute significantly to landfill waste, ocean pollution, and the degradation of ecosystems. As the world seeks sustainable solutions to this pressing issue, hemp emerges as a promising alternative for producing biodegradable plastics—offering an eco-friendly solution that can help mitigate the plastic waste crisis.

Understanding Bioplastics

Bioplastics are a type of plastic derived from renewable biological sources, such as plants, rather than fossil fuels. They can be designed to be biodegradable or compostable, which means they can break down naturally over time, reducing their impact on the environment. Hemp, with its fibrous structure and versatility, is an ideal candidate for bioplastic production.

The Advantages of Hemp-Based Bioplastics

1. Sustainability: Hemp is a fast-growing crop that requires minimal resources to cultivate. It can be harvested multiple times a year, making it a renewable source for bioplastics. In contrast to conventional plastics derived from petroleum, hemp-based bioplastics support sustainable agricultural practices.

2. Biodegradability: Hemp bioplastics can be designed to biodegrade in natural environments, significantly reducing the long-term impact of plastic waste. Unlike traditional plastics, which can take hundreds of years to decompose, hemp-based alternatives can break down within months, returning to the ecosystem without leaving harmful residues.

3. Carbon Sequestration: The growth of hemp plants captures carbon dioxide from the atmosphere, contributing to carbon sequestration. When used to produce bioplastics, the carbon remains stored in the material, further helping to mitigate climate change.

4. Reduced Dependency on Fossil Fuels: By utilizing hemp as a raw material for bioplastics, we can decrease our reliance on fossil fuels. This shift not only helps reduce greenhouse gas emissions but also promotes energy independence and sustainability.

Types of Hemp-Based Bioplastics

Hemp can be processed into various forms of bioplastics, including:

1. Hemp Fiber Composites: These bioplastics combine hemp fibers with biodegradable resins to create durable materials suitable for a range of applications, from automotive parts to consumer goods. The strength and flexibility of hemp fibers enhance the performance of these composites.

2. Hemp Seed Oil Plastics: Hemp seed oil can be processed to create bioplastics that are both biodegradable and non-toxic. These materials can be used for packaging, containers, and other products, providing a sustainable alternative to conventional plastic.

3. Hemp Starch Plastics: Hemp starch can be extracted and processed into biodegradable plastics. These materials can be used for disposable items such as cutlery and plates, offering a compostable alternative for single-use products.

Real-World Applications

Several companies and researchers are already exploring the potential of hemp-based bioplastics. Innovations in this field include biodegradable packaging solutions, compostable bags, and eco-friendly consumer products. For example, companies in the automotive industry are beginning to integrate hemp fiber composites into vehicle interiors, reducing weight and improving sustainability.

In addition, hemp-based bioplastics are finding applications in the agricultural sector, where they can be used for biodegradable mulch films and seedling pots that break down naturally in the soil, contributing to more sustainable farming practices.

Challenges and Considerations

While the potential of hemp-based bioplastics is promising, several challenges must be addressed for widespread adoption:

1. Cost and Production Scalability: The production of hemp bioplastics can be more expensive than conventional plastics due to the processing required. Developing cost-effective methods for scaling up production will be crucial to compete with traditional plastic prices.

2. Consumer Awareness: Educating consumers about the benefits of hemp-based bioplastics is essential for driving demand. As awareness of the environmental impact of plastic waste grows, consumers may increasingly seek out sustainable alternatives.

3. Regulatory Framework: Establishing regulations and standards for bioplastics, including those made from hemp, will be necessary to ensure safety and performance. Clear guidelines can help build consumer confidence and promote market growth.

Hemp-based bioplastics offer a sustainable and innovative solution to the global plastic waste crisis. By harnessing the natural properties of hemp, we can create biodegradable alternatives that reduce our reliance on fossil fuels and minimize environmental impact. As we face the challenges of plastic pollution, the potential of hemp to revolutionize the plastic industry and contribute to a circular economy becomes increasingly evident.

In the following chapters, we will explore the broader implications of hemp cultivation and its potential to address various social, economic, and environmental challenges, solidifying its role as a key player in creating a more sustainable future.

Chapter 4: Hemp in Agriculture

4.1 Crop Rotation

Crop rotation is an essential agricultural practice that involves alternating the types of crops grown in a particular field over different seasons or years. This method offers numerous benefits for soil health, pest management, and overall farm sustainability. Hemp, with its unique properties and growth characteristics, can play a vital role in crop rotation systems, enhancing agricultural productivity and promoting environmental stewardship.

The Importance of Crop Rotation

1. Pest and Disease Management: Continuous cropping of the same species can lead to the buildup of pests and diseases that target specific plants. By introducing hemp into the rotation, farmers can disrupt these cycles. Hemp's growth habits and natural resistance to certain pests can help reduce infestations and lower the need for chemical pesticides.

2. Soil Health Improvement: Different crops have varying nutrient requirements and root structures, which means they interact with the soil differently. Crop rotation helps improve soil structure, fertility, and microbial diversity. When hemp is included in the rotation, its robust root system can enhance soil aeration and promote the development of beneficial soil organisms.

3. Nutrient Management: Hemp is known as a dynamic accumulator, meaning it can extract and concentrate nutrients from the soil, particularly nitrogen. When hemp is grown, it draws up these nutrients, which can then be returned to the soil after harvest. This process helps replenish soil fertility and reduces the need for synthetic fertilizers.

4. Weed Suppression: Diverse cropping systems can suppress weed growth more effectively than monocultures. Hemp's dense canopy and rapid growth can outcompete many weeds, reducing their prevalence and the need for herbicides.

Integrating Hemp into Crop Rotation

Farmers can integrate hemp into their crop rotation systems in various ways to maximize the benefits it offers:

1. Sequential Crop Rotation: Hemp can be planted in sequence with other crops. For example, farmers may plant hemp after a cereal crop, such as wheat or corn, which can help break pest cycles and improve soil health. Following hemp with legumes can further enhance nitrogen levels in the soil.

2. Cover Cropping: Hemp can also be utilized as a cover crop during the off-season. By planting hemp in fallow periods, farmers can protect the soil from erosion, improve soil structure, and enhance nutrient cycling. Cover cropping with hemp can also promote biodiversity and support beneficial insects.

3. Intercropping: In some cases, farmers may choose to intercrop hemp with compatible crops. This practice involves growing two or more crops in close proximity to maximize land use and resource efficiency. For instance, hemp can be intercropped with legumes, which can benefit from the nitrogen-fixing properties of the legume plants while simultaneously benefiting from hemp's pest-reducing effects.

Case Studies and Success Stories

Several agricultural initiatives around the world have successfully integrated hemp into crop rotation systems. In Canada, farmers have reported improved soil health and increased yields after incorporating hemp into their rotations. By alternating hemp with traditional crops, they have seen a reduction in pest populations and a decrease in the need for chemical inputs.

In Europe, organic farmers have begun to adopt hemp as a rotational crop, citing its ability to improve soil fertility and suppress weeds. These practices have led to more sustainable farming systems that enhance both productivity and environmental health.

Challenges and Considerations

While integrating hemp into crop rotation offers numerous benefits, farmers must consider several factors for successful implementation:

1. Market Demand: Farmers need to assess the market demand for hemp and its products to ensure that their investment in hemp cultivation will be economically viable. Building local markets for hemp can help support this transition.

2. Regulatory Barriers: In some regions, regulatory restrictions on hemp cultivation may pose challenges. Farmers must navigate these regulations and advocate for supportive policies that promote hemp as a legitimate agricultural crop.

3. Education and Training: Farmers may require education and training on best practices for growing hemp and integrating it into their crop rotation systems. Access to resources and support networks can facilitate this process.

Hemp presents a valuable opportunity for farmers to enhance their crop rotation systems, improve soil health, and manage pests sustainably. By incorporating hemp into their agricultural practices, farmers can achieve better yields, reduce reliance on chemical inputs, and promote environmental stewardship.

As we continue to explore the role of hemp in agriculture, the following sections will delve into its potential for animal feed, its contributions to food systems, and its overall impact on sustainable farming practices. Together, these aspects highlight hemp's versatility and importance in creating a resilient and sustainable agricultural future.

4.2 Animal Feed

As the demand for sustainable and nutritious animal feed continues to rise, hemp seeds have emerged as a valuable resource for livestock farming. Rich in essential nutrients, hemp seeds offer a range of benefits for animal health and productivity. This chapter explores the advantages of incorporating hemp into livestock diets and its potential impact on the agriculture industry.

Nutritional Profile of Hemp Seeds

Hemp seeds are packed with essential nutrients that can significantly enhance animal diets. Key components of hemp seeds include:

1. Protein: Hemp seeds contain a high protein content, ranging from 25% to 30%. The protein is easily digestible and contains all nine essential amino acids, making it a complete protein source for animals.

2. Healthy Fats: Hemp seeds are rich in essential fatty acids, particularly omega-3 and omega-6 fatty acids, in an optimal ratio. These healthy fats contribute to overall animal health,

promoting better skin, coat quality, and reproductive performance.

3. Vitamins and Minerals: Hemp seeds are a good source of vitamins (such as vitamin E, B vitamins) and minerals (including magnesium, potassium, and phosphorus). These nutrients play vital roles in supporting immune function, bone health, and overall metabolic processes.

4. Fiber: The fibrous hulls of hemp seeds provide dietary fiber, which is important for digestive health in livestock. Fiber can enhance gut motility and contribute to a balanced diet.

Benefits of Hemp in Livestock Farming

1. Improved Animal Health: The nutrient-rich profile of hemp seeds can lead to better overall health and well-being in livestock. Animals fed diets supplemented with hemp seeds have shown improvements in growth rates, reproductive performance, and immune responses.

2. Enhanced Feed Efficiency: Research indicates that incorporating hemp seeds into animal diets can improve feed conversion ratios, meaning animals can achieve better weight

gain with less feed. This efficiency is beneficial for farmers looking to optimize production costs.

3. Sustainable Feed Source: As a renewable crop, hemp offers a sustainable alternative to traditional feed sources, such as soy and corn. Its low water requirements and ability to thrive in various soil conditions make it an environmentally friendly choice for livestock feed.

4. Reduced Use of Antibiotics: The nutritional benefits of hemp seeds can contribute to healthier livestock, potentially reducing the need for antibiotics and other veterinary interventions. This is particularly important in an era of increasing concern over antibiotic resistance.

5. Positive Impact on Meat Quality: The inclusion of hemp seeds in livestock diets can improve the fatty acid profile of meat and dairy products, resulting in healthier and more nutritious products for consumers. The omega-3 fatty acids found in hemp can enhance the nutritional quality of meat, eggs, and milk.

Applications in Different Livestock

Hemp seeds can be incorporated into the diets of various livestock, including:

1. Cattle: Hemp seeds can be fed to cattle as a supplement to enhance growth and improve the quality of beef and dairy products. The high protein content and healthy fats make them an excellent choice for both beef and dairy cattle.

2. Poultry: Chickens, ducks, and other poultry can benefit from hemp seed supplementation. Research has shown that adding hemp seeds to poultry diets can improve egg production and enhance the nutritional profile of eggs.

3. Swine: Feeding hemp seeds to pigs can lead to improved growth rates and feed efficiency. The omega fatty acids in hemp can also contribute to better meat quality.

4. Small Ruminants: Sheep and goats can benefit from the nutritional profile of hemp seeds, leading to improved health and productivity.

Challenges and Considerations

While the benefits of using hemp seeds as animal feed are clear, there are some challenges to consider:

1. Regulatory Issues: In some regions, regulations regarding hemp cultivation and its use in animal feed can pose barriers. Farmers and livestock producers need to be aware of and comply with local regulations.

2. Market Awareness: Educating livestock producers about the benefits of hemp seeds and promoting their use in animal diets is essential for fostering acceptance and adoption.

3. Cost and Availability: The cost of hemp seeds and their availability can vary based on market conditions. Ensuring a consistent supply of high-quality hemp seeds is crucial for livestock producers.

Hemp seeds present a highly nutritious and sustainable option for livestock feed, offering numerous benefits for animal health, productivity, and environmental sustainability. By incorporating hemp into animal diets, farmers can enhance growth rates, improve feed efficiency, and contribute to the production of healthier meat, dairy, and egg products.

As we continue to explore the role of hemp in agriculture, the next sections will address its contributions to food systems and its broader impact on sustainable farming practices. Together, these aspects highlight hemp's versatility and importance in creating a resilient agricultural future.

Chapter 5: Hemp for Human Health

5.1 Nutritional Benefits

Hemp seeds, often referred to as "superfoods," have garnered significant attention for their impressive nutritional profile. Packed with essential fatty acids, high-quality protein, vitamins, and minerals, hemp seeds offer a range of health benefits that can contribute to overall well-being. This chapter explores the various nutritional benefits of incorporating hemp into our diets and its potential role in promoting a healthy lifestyle.

Nutritional Profile of Hemp Seeds

1. Essential Fatty Acids: Hemp seeds are an excellent source of polyunsaturated fatty acids, particularly omega-3 (alpha-linolenic acid, ALA) and omega-6 (linoleic acid) fatty acids. The ideal ratio of omega-3 to omega-6 in hemp seeds (approximately 1:3) is beneficial for heart health, reducing inflammation, and supporting brain function.

2. High-Quality Protein: Hemp seeds contain about 25% to 30% protein, making them one of the best plant-based protein sources. They provide all nine essential amino acids, which are crucial for muscle repair, immune function, and overall health. This makes hemp seeds especially valuable for vegetarians, vegans, and those looking to reduce their meat consumption.

3. Vitamins and Minerals: Hemp seeds are rich in essential vitamins and minerals, including:

   - Vitamin E: A powerful antioxidant that helps protect cells from oxidative stress and supports skin health.

   - B Vitamins: Including B1 (thiamine), B2 (riboflavin), B3 (niacin), and B6, which play vital roles in energy metabolism, brain function, and red blood cell formation.

   - Magnesium: Important for muscle function, nerve signaling, and bone health.

   - Zinc: Essential for immune function, protein synthesis, and wound healing.

4. Dietary Fiber: Although hemp seeds are primarily known for their nutritional content, they also contain dietary fiber, particularly in their hulls. Fiber is important for digestive health, promoting regular bowel movements, and supporting gut microbiota.

Health Benefits of Incorporating Hemp into Your Diet

1. Heart Health: The presence of omega-3 and omega-6 fatty acids in hemp seeds can help lower cholesterol levels, reduce blood pressure, and improve overall cardiovascular health. These fatty acids also play a role in reducing inflammation, which is linked to heart disease.

2. Weight Management: Hemp seeds can support weight management due to their high protein content and healthy fats, which promote satiety and reduce hunger. Including hemp seeds in meals can help control cravings and support healthy eating patterns.

3. Muscle Repair and Recovery: The complete protein profile of hemp seeds makes them an excellent choice for athletes and active individuals. Consuming hemp seeds post-workout can aid in muscle repair and recovery, supporting overall physical performance.

4. Skin Health: The fatty acids and antioxidants found in hemp seeds can benefit skin health by promoting hydration and

reducing inflammation. Some studies suggest that hemp oil may help alleviate skin conditions such as eczema and psoriasis.

5. Hormonal Balance: The gamma-linolenic acid (GLA) found in hemp seeds is known to support hormonal balance, particularly in women. GLA may help alleviate symptoms associated with premenstrual syndrome (PMS) and menopause.

6. Digestive Health: The fiber content in hemp seeds supports digestive health by promoting regular bowel movements and feeding beneficial gut bacteria. A healthy gut microbiome is essential for overall health and well-being.

Ways to Incorporate Hemp into Your Diet

Hemp seeds are versatile and can be easily incorporated into various meals and snacks. Here are some popular ways to enjoy hemp seeds:

1. Smoothies: Add a tablespoon or two of hemp seeds to smoothies for a protein and nutrient boost.

2. Salads: Sprinkle hemp seeds on salads for added crunch and nutrition.

3. Baking: Incorporate hemp seeds into baked goods like muffins, bread, and granola bars.

4. Oatmeal and Yogurt: Mix hemp seeds into oatmeal or yogurt for an easy and nutritious breakfast.

5. Energy Balls: Create homemade energy balls or protein bars using hemp seeds, nut butter, and other nutritious ingredients.

Hemp seeds are a powerhouse of nutrition, offering a wide array of health benefits. Their rich content of essential fatty acids, high-quality protein, vitamins, and minerals make them an excellent addition to any diet. By incorporating hemp seeds into our meals, we can promote heart health, support muscle recovery, and enhance overall well-being.

As we continue to explore the role of hemp in nutrition, the next sections will delve into other aspects of hemp's contributions to food systems and its broader impact on sustainable dietary practices. Together, these elements underscore hemp's potential as a valuable resource for healthier living and a more sustainable future.

5.2 Medicinal Uses

Hemp has been utilized for its medicinal properties for centuries, but in recent years, the spotlight has largely fallen on hemp-derived cannabidiol (CBD). This non-psychoactive compound, along with other cannabinoids found in hemp, has gained significant popularity for its potential health benefits. This chapter explores the medicinal properties of hemp and the various ways it can contribute to health and wellness.

Understanding CBD and Its Properties

Cannabidiol (CBD) is one of over 100 cannabinoids found in the hemp plant. Unlike tetrahydrocannabinol (THC), the primary psychoactive compound in cannabis, CBD does not produce a "high." Instead, it interacts with the body's endocannabinoid system (ECS), which plays a vital role in regulating various physiological processes, including mood, pain perception, immune response, and sleep.

Potential Health Benefits of Hemp-Derived CBD

1. Pain Relief: One of the most well-documented uses of CBD is its potential to alleviate chronic pain. Research suggests that CBD may help reduce inflammation and interact with neurotransmitters to modulate pain signaling. This has made CBD an appealing option for individuals suffering from conditions such as arthritis, multiple sclerosis, and fibromyalgia.

2. Anxiety and Stress Reduction: Numerous studies have indicated that CBD may help reduce anxiety and stress levels. By interacting with receptors in the brain associated with mood regulation, CBD may promote a sense of calm and relaxation. Some research suggests that it may be beneficial for individuals with generalized anxiety disorder, social anxiety disorder, and post-traumatic stress disorder (PTSD).

3. Sleep Improvement: CBD has gained attention for its potential to improve sleep quality. By addressing issues related to anxiety and pain, CBD may help individuals fall asleep more easily and stay asleep longer. Some studies have shown that CBD can help regulate sleep patterns, making it a potential option for those dealing with insomnia.

4. Neuroprotective Properties: Preliminary research suggests that CBD may have neuroprotective effects, which could be beneficial for individuals with neurodegenerative diseases such

as Alzheimer's and Parkinson's. CBD's anti-inflammatory and antioxidant properties may help protect brain cells from damage.

5. Anti-Seizure Effects: CBD has been recognized for its potential anti-seizure properties, particularly in the treatment of epilepsy. The FDA has approved Epidiolex, a CBD-based medication, for the treatment of certain forms of epilepsy, highlighting its effectiveness in reducing seizure frequency in some patients.

6. Skin Health: CBD's anti-inflammatory properties may also benefit skin health. Studies have shown that CBD can help regulate oil production in the skin and may be effective in treating conditions like acne and eczema. Topical CBD products are becoming increasingly popular for their potential to soothe irritated skin.

Methods of Consumption

There are several ways to consume hemp-derived CBD, each with its own advantages:

1. Oils and Tinctures: CBD oils and tinctures are concentrated extracts that can be taken sublingually (under the tongue) for quick absorption.

2. Capsules and Edibles: CBD capsules and edibles (like gummies) offer a convenient and discreet way to consume CBD. However, they may take longer to take effect as they need to be digested.

3. Topicals: CBD-infused creams, balms, and lotions can be applied directly to the skin for localized relief of pain and inflammation.

4. Vaping: Some individuals prefer using vape pens or CBD cartridges for quick absorption. However, the long-term safety of vaping is still under investigation.

5. Beverages: CBD-infused beverages, such as tea or sparkling water, are gaining popularity as a refreshing way to consume CBD.

Challenges and Considerations

While the potential health benefits of hemp-derived CBD are promising, there are important considerations to keep in mind:

1. Regulation and Quality Control: The CBD market is largely unregulated, leading to variations in product quality and potency. It's essential for consumers to choose products from reputable manufacturers that provide third-party lab testing to ensure safety and efficacy.

2. Individual Responses: The effects of CBD can vary from person to person. Factors such as dosage, method of consumption, and individual biochemistry can influence how someone responds to CBD.

3. Legal Status: The legal status of hemp-derived CBD varies by region. While the 2018 Farm Bill legalized hemp cultivation and the production of CBD in the United States, local regulations may still apply.

4. Potential Side Effects: While CBD is generally well-tolerated, some individuals may experience side effects such as fatigue, changes in appetite, or gastrointestinal discomfort. It's advisable to consult with a healthcare professional before starting any new

supplement, especially for individuals with pre-existing medical conditions or those taking other medications.

Hemp-derived CBD holds significant promise as a natural remedy for various health issues, including pain relief, anxiety reduction, and improved sleep quality. As research continues to uncover the full potential of CBD and other hemp compounds, individuals seeking alternative health solutions may find valuable options in hemp products.

In the following sections, we will delve deeper into the broader implications of hemp in food systems and its potential to contribute to sustainable dietary practices, illustrating its role as a versatile and beneficial resource for a healthier future.

Chapter 6: Economic Opportunities

6.1 Job Creation

As the hemp industry continues to gain momentum globally, its potential for job creation becomes increasingly evident. From farming and processing to manufacturing and retail, the diverse applications of hemp can generate numerous employment opportunities across various sectors. This chapter examines the economic benefits of a thriving hemp industry and its capacity to contribute to local and national economies.

The Expanding Hemp Market

The resurgence of hemp cultivation has been fueled by growing consumer demand for hemp-based products, including textiles, bioplastics, food, cosmetics, and health supplements like CBD. As industries seek sustainable alternatives to traditional materials, the hemp market is poised for significant growth. This expansion not only creates jobs within the industry but also stimulates economic activity in related sectors.

Job Opportunities in the Hemp Industry

1. Farming and Cultivation: As more farmers adopt hemp as a viable crop, there is an increased demand for labor in planting, harvesting, and managing hemp fields. This includes positions for farm workers, agronomists, and agricultural technicians who specialize in hemp cultivation practices.

2. Processing and Manufacturing: After harvesting, hemp requires processing to extract fibers, seeds, and oils. The processing sector encompasses a wide range of jobs, including machine operators, quality control specialists, and facility managers. Additionally, manufacturers that produce hemp-based products—such as textiles, paper, construction materials, and bioplastics—also contribute significantly to job creation.

3. Research and Development: The growing interest in hemp has led to increased investment in research and development (R&D) to explore its potential applications and improve cultivation practices. This creates jobs for scientists, researchers, and engineers focused on innovation in hemp technology, agriculture, and product development.

4. Distribution and Retail: As the market for hemp products expands, opportunities in distribution, logistics, and retail emerge. This includes positions in warehousing, supply chain management, and retail sales for hemp products in health food stores, specialty shops, and online platforms.

5. Education and Advocacy: The growing hemp industry also requires professionals involved in education, training, and advocacy. This includes agricultural extension agents, educators, and non-profit organizations that promote hemp cultivation and its benefits, creating jobs that support community engagement and awareness.

Economic Impact on Local Communities

The economic benefits of a thriving hemp industry extend beyond job creation. The cultivation and processing of hemp can have a positive ripple effect on local economies:

1. Increased Revenue: Farmers growing hemp can diversify their income sources, leading to increased revenue for agricultural communities. This additional income can support local businesses and stimulate economic growth.

2. Support for Ancillary Industries: The growth of the hemp industry can benefit related sectors, including transportation, packaging, and retail. As demand for hemp products increases, businesses in these areas may expand, further contributing to job creation.

3. Investment in Infrastructure: The development of the hemp industry can lead to investments in infrastructure, such as processing facilities, transportation networks, and research institutions. These investments can improve local economies and create additional employment opportunities.

4. Rural Revitalization: Hemp cultivation can help revitalize rural areas facing economic challenges. By providing new opportunities for farmers and creating jobs in processing and manufacturing, the hemp industry can contribute to the sustainability of rural communities.

Case Studies and Success Stories

Several regions around the world have already experienced the economic benefits of a thriving hemp industry:

- United States: Following the legalization of hemp cultivation under the 2018 Farm Bill, many states have seen a surge in hemp farming and related businesses. States like Colorado and Kentucky have established robust hemp industries, creating thousands of jobs and stimulating local economies.

- Canada: Canada has been a leader in hemp cultivation and processing for decades. The Canadian hemp industry has created jobs in farming, processing, and product manufacturing, contributing significantly to rural economies.

- Europe: Countries such as France, Germany, and the Netherlands have recognized the economic potential of hemp and are investing in its cultivation and processing. These nations have developed strong hemp markets, leading to job creation in agriculture, manufacturing, and research.

Challenges and Considerations

While the hemp industry offers considerable job creation potential, several challenges must be addressed:

1. Regulatory Framework: The development of a clear regulatory framework is essential for the growth of the hemp industry. Uncertainty regarding cultivation, processing, and marketing regulations can hinder investment and job creation.

2. Market Stability: The hemp market can be volatile, influenced by fluctuations in demand and pricing. Ensuring a stable market for hemp products is crucial for sustaining job growth and economic impact.

3. Education and Training: As the hemp industry evolves, there is a need for education and training programs to equip workers with the necessary skills. Investment in workforce development is essential to support the growth of the industry.

The hemp industry holds significant promise for job creation and economic revitalization. As demand for hemp-based products continues to rise, the potential for employment across various sectors—from farming to manufacturing—will expand, benefiting local and national economies. By fostering a thriving hemp industry, we can create sustainable job opportunities while promoting environmental stewardship and innovation.

In the subsequent sections, we will explore further economic impacts of hemp, including its potential for export opportunities and its role in fostering sustainable business practices, underscoring hemp's significance as a valuable resource for a resilient economic future.

6.2 Global Trade

As the global demand for sustainable and eco-friendly products continues to rise, hemp emerges as a valuable commodity with significant export potential. Countries that invest in hemp cultivation and processing can diversify their economies, create jobs, and tap into the growing international market for hemp-based products. This chapter explores the global market potential for hemp and the opportunities it presents for trade and economic development.

The Growing Global Market for Hemp

The global hemp market has experienced remarkable growth in recent years, driven by increasing consumer awareness of environmental issues and the search for sustainable alternatives to traditional materials. Key factors contributing to the expanding market include:

1. Consumer Demand: There is a growing consumer preference for hemp-based products, ranging from textiles and biodegradable plastics to food, supplements, and personal care

items. As more consumers seek eco-friendly options, the demand for hemp is expected to continue rising.

2. Legislative Changes: The legalization of hemp cultivation in various countries has opened new markets and opportunities for trade. The 2018 Farm Bill in the United States and similar legislative changes in Canada and Europe have created a more favorable environment for hemp production and export.

3. Sustainability Goals: Many countries are striving to meet sustainability goals and reduce their reliance on fossil fuels and non-renewable resources. Hemp's versatility and environmentally friendly attributes make it an attractive option for industries seeking sustainable materials.

Key Hemp Products in Global Trade

Hemp can be processed into a wide variety of products, each with its own market potential:

1. Textiles: Hemp fibers are highly durable and biodegradable, making them an excellent alternative to cotton and synthetic fibers. The global demand for sustainable textiles presents significant export opportunities for countries producing hemp-based fabrics.

2. Bioplastics: With the growing focus on reducing plastic pollution, hemp-derived bioplastics are gaining traction in various industries, including packaging and automotive. Countries that develop bioplastic manufacturing capabilities using hemp can tap into this expanding market.

3. Food and Supplements: Hemp seeds and oil are rich in nutrients and are increasingly popular in health food markets. The demand for hemp-based food products, including protein powders and health supplements, is growing in both domestic and international markets.

4. CBD Products: The global market for CBD (cannabidiol) is one of the fastest-growing segments in the hemp industry. As more countries legalize CBD for therapeutic use, there is substantial potential for export, particularly for high-quality CBD oils and products.

5. Construction Materials: Hempcrete and other hemp-based building materials are gaining recognition in the construction industry for their sustainability and performance benefits. Countries that innovate in this area can export these materials to meet the demand for eco-friendly construction solutions.

Opportunities for Economic Diversification

For countries looking to diversify their economies, hemp presents a compelling opportunity:

1. Rural Development: Hemp cultivation can revitalize rural communities by providing farmers with a new crop, increasing agricultural diversity, and creating jobs in processing and manufacturing. This can help sustain rural economies and reduce urban migration.

2. Innovation and Research: Investment in hemp research and development can lead to innovations in cultivation techniques, processing methods, and product applications. Countries that prioritize R&D in hemp can position themselves as leaders in the global market.

3. Export Revenue: By developing a robust hemp industry, countries can generate significant export revenue. This revenue can be reinvested in local communities and infrastructure, further stimulating economic growth.

4. Collaborative Trade Agreements: Countries can forge trade agreements that facilitate the exchange of hemp products, enhancing market access and creating partnerships that benefit both producers and consumers.

Challenges and Considerations

Despite the promising potential of the global hemp market, several challenges must be addressed:

1. Regulatory Hurdles: The legal status of hemp varies across countries, leading to complexities in international trade. Streamlining regulations and harmonizing standards can help facilitate trade.

2. Market Competition: As the global hemp market grows, competition will increase. Countries must focus on quality, innovation, and sustainable practices to remain competitive.

3. Infrastructure Development: Adequate infrastructure for processing, transportation, and distribution is essential for maximizing the economic benefits of hemp trade. Investment in infrastructure will be necessary to support a thriving hemp industry.

4. Consumer Education: Educating consumers about the benefits of hemp products is crucial for driving demand. Raising

awareness and promoting the advantages of hemp can help expand market opportunities.

Hemp has significant potential as a valuable export commodity for countries seeking to diversify their economies and tap into the growing global market for sustainable products. By investing in hemp cultivation, processing, and innovation, countries can create jobs, stimulate economic growth, and contribute to global sustainability efforts.

As we continue to explore the economic impacts of hemp, the next sections will delve into its potential for fostering sustainable business practices and its broader implications for economic resilience, highlighting hemp's role as a cornerstone of a more sustainable and prosperous future.

Chapter 7: Policy and Regulation

7.1 Legal Challenges

Despite the numerous benefits associated with hemp cultivation, including its potential for economic growth, environmental sustainability, and health benefits, the legal landscape surrounding hemp remains complex and often restrictive. Many countries still impose significant barriers to hemp farming, processing, and trade. This chapter examines the legal hurdles that hinder the growth of the hemp industry and the pressing need for policy reform.

Historical Context

The legal challenges surrounding hemp cultivation primarily stem from historical misconceptions and stigma associated with cannabis. For much of the 20th century, hemp was lumped together with marijuana, resulting in strict regulations and prohibitions on its cultivation. In many countries, including the United States, hemp was classified as a Schedule I controlled substance under the Controlled Substances Act, effectively criminalizing its production.

However, as scientific research has highlighted the distinct differences between hemp and marijuana, a growing awareness of hemp's benefits has emerged. In recent years, several countries have begun to reform their policies to facilitate hemp cultivation, but significant legal challenges remain.

Current Legal Challenges

1. Complex Regulations: In many regions, the legal framework governing hemp is convoluted and varies significantly from one jurisdiction to another. This complexity can create confusion for farmers, processors, and businesses looking to enter the hemp market. Navigating the regulatory landscape may require significant legal and administrative resources.

2. Licensing Requirements: In many countries, hemp farmers must obtain specific licenses to cultivate hemp, which can be time-consuming and costly. These licensing processes often involve extensive paperwork, background checks, and compliance with various agricultural regulations. The burden of

these requirements can deter potential farmers from entering the hemp industry.

3. THC Limits: Most jurisdictions impose strict limits on the allowable concentration of tetrahydrocannabinol (THC), the psychoactive compound found in cannabis, in hemp plants. While these limits are intended to differentiate hemp from marijuana, they can create challenges for farmers. Variability in THC levels due to environmental factors, genetics, and cultivation practices can lead to unexpected compliance issues and crop loss.

4. Interstate and International Trade Barriers: In countries with federal regulations, such as the United States, the legal status of hemp can complicate interstate trade. Conflicting state laws can create uncertainty for businesses operating across state lines. Additionally, international trade of hemp products may be hampered by differing regulations and tariffs.

5. Lack of Research and Development Support: Research into the benefits and applications of hemp is often limited due to funding restrictions and legal barriers. The absence of robust research can hinder innovation in cultivation techniques, product development, and market expansion, ultimately stunting the growth of the hemp industry.

The Need for Policy Reform

To unlock the full potential of the hemp industry, comprehensive policy reform is essential. Key areas for reform include:

1. Simplifying Regulations: Streamlining the regulatory framework governing hemp cultivation and processing can reduce confusion and make it easier for farmers and businesses to participate in the market. Clear and consistent regulations are crucial for fostering a thriving hemp industry.

2. Establishing Supportive Licensing Processes: Policymakers should aim to create efficient and accessible licensing processes for hemp farmers. This includes reducing bureaucratic hurdles and providing clear guidelines for compliance.

3. Raising THC Limits Creatively: Reevaluating and potentially raising THC limits for hemp cultivation can help accommodate the natural variability in cannabis plants. Implementing more flexible testing protocols and allowing for a buffer zone can reduce compliance issues for farmers.

4. Promoting Research and Development: Increased investment in hemp research and development is crucial for promoting innovation and exploring the full range of hemp's applications. Governments can support research initiatives through grants, partnerships with universities, and incentives for private sector investment.

5. Encouraging International Cooperation: Countries should work together to harmonize regulations related to hemp cultivation and trade. International cooperation can facilitate market access and promote fair trade practices, benefiting producers and consumers alike.

Case Studies of Successful Legal Reforms

Several countries and regions have successfully reformed their hemp policies, serving as examples of how legal barriers can be overcome:

- United States: The passage of the 2018 Farm Bill marked a significant turning point for hemp in the U.S. by legalizing its cultivation and removing it from the list of controlled substances. States have since developed their own regulatory frameworks, enabling a resurgence of the hemp industry.

- Canada: Canada has long been a leader in hemp cultivation, having legalized hemp production in 1998. The Canadian government continues to support the industry through research funding and market development initiatives, contributing to its growth.

- European Union: The EU has established regulations that allow member states to cultivate hemp varieties with low THC content. This framework has facilitated hemp production across Europe, leading to increased market opportunities for farmers and businesses.

The legal challenges surrounding hemp cultivation remain significant barriers to the industry's growth. However, the momentum for reform is building as awareness of hemp's benefits continues to grow. By addressing these legal hurdles through comprehensive policy reform, governments can unlock the potential of the hemp industry, fostering economic development, sustainability, and innovation.

In the subsequent chapters, we will further explore the social implications of hemp cultivation and its potential to contribute to sustainable practices in various sectors, reinforcing hemp's role as a valuable resource in the modern economy.

7.2 Advocacy and Education

Public perception of hemp and its various applications plays a crucial role in shaping policy and legislation surrounding its cultivation and use. As awareness of the benefits of hemp continues to grow, advocacy and educational efforts are essential in transforming misconceptions and promoting informed decision-making among both the public and policymakers. This chapter explores the initiatives aimed at educating people about the benefits of hemp and the importance of advocacy in driving positive change.

The Importance of Advocacy

Advocacy is the act of supporting a cause or proposal and can take many forms, including grassroots organizing, lobbying, public campaigns, and community engagement. In the context of hemp, effective advocacy is vital for several reasons:

1. Changing Perceptions: Misunderstandings and stigmas associated with hemp often stem from its historical association with marijuana. Advocacy efforts aim to educate the public

about the differences between the two, emphasizing that hemp is a versatile, non-psychoactive plant with numerous benefits.

2. Influencing Policy: Advocates play a critical role in shaping policy by providing lawmakers with data, research, and real-life stories that illustrate the benefits of hemp. Informed policymakers are more likely to support legislation that promotes hemp cultivation and removes unnecessary barriers.

3. Building Community Support: Advocacy efforts help build community support for hemp initiatives, creating a network of stakeholders, including farmers, businesses, researchers, and consumers. This collective voice can amplify the call for change and encourage local governments to adopt supportive policies.

Educational Initiatives

Education is a fundamental component of advocacy efforts. Various organizations, institutions, and individuals are working to increase awareness and understanding of hemp through educational initiatives:

1. Workshops and Seminars: Many advocacy groups host workshops and seminars to educate farmers, entrepreneurs, and consumers about the benefits of hemp and its potential applications. These events often cover topics such as cultivation techniques, processing methods, and product development.

2. Public Campaigns: Advocacy organizations run public awareness campaigns to inform the general public about the benefits of hemp. These campaigns may utilize social media, traditional advertising, and community events to reach a wide audience and dispel myths about hemp.

3. Research and Publications: Academic institutions and research organizations conduct studies on the economic, environmental, and health benefits of hemp, contributing valuable data to support advocacy efforts. Publications, reports,

and white papers can serve as important resources for policymakers and the public.

4. School and Community Programs: Some educational initiatives focus on integrating hemp education into school curricula and community programs. By teaching younger generations about hemp's potential, advocates can cultivate a more informed future workforce and consumer base.

5. Partnerships with Industry Leaders: Collaborating with industry leaders and businesses can enhance educational efforts by providing real-world insights and expertise. These partnerships can facilitate knowledge sharing and promote best practices in hemp cultivation and processing.

Successful Advocacy Examples

Several successful advocacy efforts have made significant strides in changing public perception and influencing policy regarding hemp:

- Hemp Industries Association (HIA): This organization has been at the forefront of hemp advocacy in the United States, working to promote hemp education, support legislative efforts, and connect stakeholders in the hemp industry. Their efforts have contributed to the passage of laws that support hemp cultivation and commerce.

- Project CBD: Focusing specifically on the health benefits of cannabidiol (CBD) derived from hemp, Project CBD provides educational resources and advocates for policies that support CBD research and accessibility. Their work has helped demystify CBD and promote its therapeutic potential.

- Global Hemp Initiative: This international coalition brings together advocates, researchers, and industry leaders to promote hemp as a sustainable resource. Through educational campaigns and collaborative efforts, they aim to raise awareness of hemp's benefits on a global scale.

Challenges to Advocacy

While advocacy efforts are essential for advancing the hemp industry, several challenges remain:

1. Resistance to Change: Despite growing awareness, some policymakers and segments of the public may remain resistant to changing their views on hemp. Overcoming deeply rooted misconceptions requires persistent education and engagement.

2. Limited Funding: Advocacy organizations often rely on donations and grants to fund their initiatives. Limited financial resources can restrict their ability to reach wider audiences and implement comprehensive educational programs.

3. Fragmented Efforts: The hemp advocacy landscape can be fragmented, with various organizations pursuing different goals and strategies. Coordinating efforts and fostering collaboration among stakeholders can enhance the effectiveness of advocacy initiatives.

Advocacy and education are critical components in the journey toward a more favorable legal landscape for hemp cultivation and use. By changing public perception, influencing policy, and fostering community support, advocacy efforts can unlock the full potential of the hemp industry.

As we continue to explore the implications of hemp cultivation, the next sections will focus on the social and environmental benefits of hemp and its potential role in promoting sustainable practices, reinforcing the importance of a holistic approach to hemp as a valuable resource in the modern economy.

Increasing Awareness Through Collective Action

As we navigate the evolving landscape of hemp cultivation and its multifaceted benefits, it is clear that raising awareness is a shared responsibility. The potential of hemp to contribute to sustainable practices, economic growth, and improved health outcomes is vast, but it requires a collective effort to educate and inform those around us.

Everyone has a role to play in this movement—whether you are a farmer, a business owner, a consumer, or simply an advocate for environmental sustainability. By working together, we can change perceptions, influence policy, and foster a more informed community that recognizes the value of hemp.

One Effective Strategy: Sharing Knowledge Through This Book

One impactful way to promote awareness is by purchasing copies of this book and sharing them with friends, family, colleagues, and community members. This book serves as a comprehensive resource that outlines the benefits of hemp across various sectors, including agriculture, nutrition, medicine, and the economy. By distributing this knowledge, we can spark conversations and encourage others to explore the potential of hemp in their own lives.

Expected Impact of Sharing This Book

1. Education and Understanding: Providing copies of this book will help demystify hemp and its uses, allowing individuals to understand its potential benefits and applications. The more people know about hemp, the more likely they are to embrace it as a sustainable resource.

2. Informed Discussions: Sharing this book can catalyze discussions within families, workplaces, and communities. These conversations can lead to increased interest in hemp products and practices, fostering a culture of sustainability and innovation.

3. Advocacy and Support: As people become more informed about the benefits of hemp, they may be inspired to advocate for supportive policies and practices. This grassroots approach can influence local and national decision-makers to prioritize hemp cultivation and remove barriers to its growth.

4. Community Engagement: Distributing the book can encourage community engagement and collaboration among stakeholders, including farmers, manufacturers, and consumers.

By creating a network of informed individuals, we can collectively work toward building a robust hemp industry.

5. Economic Growth: Increased awareness and interest in hemp can lead to greater demand for hemp products, stimulating the economy and creating job opportunities within local communities. This ripple effect can have lasting benefits for both individuals and the broader economy.

We all have a vital role in raising awareness about the potential of hemp. By purchasing and sharing this book, we can educate those we care about and foster a community that values sustainability, health, and innovation. Together, we can create a more informed society that embraces the benefits of hemp and supports its growth as a key player in building a sustainable future.

Let's work together to spread the word, inspire change, and unlock the full potential of hemp for generations to come.

Hemp is a practical solution

Hemp is not a panacea, but it offers practical solutions to many of the environmental, economic, and social challenges we face today. Its versatility as a crop allows it to contribute to a wide array of sectors, from agriculture and textiles to nutrition and medicine. As we grapple with pressing issues such as climate change, resource depletion, and economic inequality, hemp presents an opportunity to rethink our approaches and embrace more sustainable practices.

By adopting hemp cultivation and utilizing its products, we can reduce our reliance on fossil fuels, promote regenerative agricultural practices, and create a circular economy that prioritizes sustainability. Hemp's potential to sequester carbon, improve soil health, and reduce waste positions it as a valuable ally in our efforts to combat environmental degradation.

Economically, the growth of the hemp industry can lead to job creation and diversification of agricultural practices, providing farmers and communities with new income streams. It can stimulate local economies and foster innovation in product development, from biodegradable materials to health supplements.

Socially, the embrace of hemp can help challenge outdated perceptions and stigmas, paving the way for more informed discussions about its benefits. By advocating for policy reforms and supporting educational initiatives, we can build a more equitable framework that recognizes the value of hemp in promoting health and well-being.

In conclusion, by embracing this versatile plant, we can take a significant step toward a more sustainable and equitable world. It is essential for individuals, communities, and policymakers to come together in support of hemp cultivation and its myriad applications. Through collective action, education, and advocacy, we can unlock the full potential of hemp and harness its benefits for a healthier planet and a thriving economy. Let us move forward with the understanding that while hemp may not solve all our problems, it certainly provides a pathway toward a brighter, more sustainable future.

--- References

Here is a comprehensive list of studies, articles, and other resources for further reading on the various aspects of hemp, including its agricultural, nutritional, medicinal, economic, and legal implications:

Books

1. Hemp Bound: Dispatches from the Front Lines of the Next Agricultural Revolution by Doug Fine.

- A detailed exploration of the hemp industry and its potential to transform agriculture and economies.

2. The Hempcrete Book: Designing and Building with Hemp-Lime by William Stanwix and Alex Sparrow.

   - A practical guide to using hempcrete as a sustainable building material.

3. The Cannabis Encyclopedia: The Definitive Guide to Cultivation & Consumption of Medical Cannabis by Jorge Cervantes.

   - While focusing on cannabis, this book provides valuable insights into the cultivation and uses of hemp.

Articles and Journals

4. "The Role of Hemp in Sustainable Agriculture" by The European Industrial Hemp Association. (2020)

   - Discusses the environmental benefits of hemp cultivation and its potential role in sustainable agriculture.

   - Link: European Industrial Hemp Association (https://eiha.org)

5. "Hemp Seed: Nutritional Profile and Health Benefits" in Nutrients Journal. (2019)

   - A peer-reviewed article that examines the nutritional benefits of hemp seeds.

   - DOI: 10.3390/nu11061305 (https://doi.org/10.3390/nu11061305)

6. "Cannabidiol (CBD) in the Treatment of Epilepsy: A Review" in Journal of Epilepsy Research. (2019)

   - A comprehensive review of the therapeutic potential of CBD for epilepsy.

   - DOI: 10.14581/jer.19006 (https://doi.org/10.14581/jer.19006)

7. "The Economic Impact of Legalizing Hemp" by the National Hemp Association. (2021)

   - An analysis of the economic benefits resulting from the legalization of hemp in the United States.

   - Link: National Hemp Association (https://nationalhempassociation.org)

8. "Hemp: A New Crop with New Uses for North America" in New Crops. (1996) by J. B. McPartland, R. C. Clarke, and D. P. Watson.

- An early comprehensive overview of hemp's potential uses and benefits in North America.

Reports and White Papers

9. "Hemp: A Sustainable Agricultural Alternative" by the United Nations. (2020)

   - A report discussing hemp's role in sustainable agriculture and its potential to address environmental challenges.

   - Link: UN Report (https://www.un.org)

10. "The Global Cannabis Report: Market Trends and Opportunities" by Prohibition Partners. (2021)

    - An extensive report on the cannabis market, including hemp, with insights into trends and economic opportunities.

Websites and Online Resources

11. Hemp Industries Association (HIA)

    - An organization dedicated to promoting the hemp industry through education, advocacy, and networking.

    - Link: Hemp Industries Association (https://www.thehia.org)

12. Project CBD

- A nonprofit dedicated to promoting and publicizing research on the medical uses of cannabidiol (CBD) and other components of cannabis.

- Link: Project CBD (https://www.projectcbd.org)

13. European Industrial Hemp Association (EIHA)

- Provides resources and information on hemp cultivation, processing, and market development in Europe.

- Link: European Industrial Hemp Association (https://eiha.org)

14. National Hemp Growers Cooperative

- A cooperative aimed at supporting hemp farmers and promoting sustainable practices.

- Link: National Hemp Growers Cooperative (https://www.nationalhempgrowerscooperative.com)

These references provide a wealth of information for those interested in exploring the diverse aspects of hemp further. Whether you're looking for in-depth studies, practical guides, or insights into the economic potential of hemp, these resources will serve as valuable tools for understanding this remarkable plant and its potential to contribute to a sustainable future.

Appendix

This appendix provides additional information on hemp, including recipes, DIY projects, and a list of organizations and businesses that focus on hemp-related products and initiatives. These resources aim to inspire you to explore the versatility of hemp and engage with the community surrounding this remarkable plant.

Recipes Using Hemp

1. Hemp Seed Smoothie
   - Ingredients:
     - 2 tablespoons hemp seeds
     - 1 banana
     - 1 cup spinach
     - 1 cup almond milk (or other milk of choice)
     - 1 tablespoon nut butter (optional)
   - Instructions:
     1. Combine all ingredients in a blender.

2. Blend until smooth.

3. Enjoy as a nutritious breakfast or snack!

2. Hemp-Infused Salad Dressing
   - Ingredients:
     - 1/4 cup hemp seed oil
     - 2 tablespoons apple cider vinegar
     - 1 teaspoon Dijon mustard
     - Salt and pepper to taste
   - Instructions:
     1. Whisk together all ingredients until well combined.
     2. Drizzle over your favorite salad and enjoy!

3. Hemp Protein Energy Balls
   - Ingredients:
     - 1 cup rolled oats
     - 1/2 cup almond butter
     - 1/4 cup honey or maple syrup
     - 1/4 cup hemp seeds
     - 1/4 cup dark chocolate chips (optional)

- Instructions:

    1. In a bowl, mix all ingredients until combined.

    2. Roll the mixture into small balls.

    3. Refrigerate for about 30 minutes before serving.

DIY Projects with Hemp

1. Hemp Cord Bracelets

    - Materials Needed:

        - Hemp cord

        - Scissors

        - Beads (optional)

    - Instructions:

        1. Cut a length of hemp cord to your desired bracelet size.

        2. If using beads, thread them onto the cord.

        3. Braid or knot the cord together to create a bracelet.

        4. Tie off the ends and wear your handmade accessory!

2. Hemp-Dyed Fabric

    - Materials Needed:

- Natural fabric (cotton, linen, etc.)
- Hemp flowers or leaves (for dye)
- Water
- Instructions:
    1. Boil hemp flowers or leaves in water to extract the dye.
    2. Strain the mixture and add your fabric to the dye bath.
    3. Simmer for 30-60 minutes, then rinse and dry.
    4. Enjoy your unique, hemp-dyed fabric!

Hemp-Related Organizations and Businesses

Organizations

1. Hemp Industries Association (HIA)

- A trade association that advocates for the hemp industry and promotes hemp education.
- Link: Hemp Industries Association (https://www.thehia.org)

2. Project CBD

- A nonprofit organization focused on promoting research and education about CBD and its potential health benefits.

- Link: Project CBD (https://www.projectcbd.org)

3. European Industrial Hemp Association (EIHA)

   - An organization promoting the interests of the hemp industry in Europe.

   - Link: EIHA (https://eiha.org)

4. National Hemp Growers Cooperative

   - A cooperative supporting hemp farmers and promoting sustainable practices.

   - Link: National Hemp Growers Cooperative (https://www.nationalhempgrowerscooperative.com)

Businesses

1. Nutiva

   - A supplier of organic hemp seeds, oils, and protein powders.

   - Link: Nutiva (https://www.nutiva.com)

2. Manitoba Harvest

- A leading hemp food company offering a variety of hemp-based products, including seeds, protein powder, and oil.

   - Link: Manitoba Harvest (https://www.manitobaharvest.com)

3. Hempitecture

   - A company specializing in sustainable building materials made from hemp.

   - Link: Hempitecture (https://www.hempitecture.com)

4. HempWorx

   - A brand offering a wide range of CBD products derived from hemp.

   - Link: HempWorx (https://hempworx.com)

This appendix serves as a starting point for exploring the diverse applications and benefits of hemp. Whether through trying out recipes, engaging in DIY projects, or connecting with organizations and businesses, there are countless ways to incorporate hemp into your life and support the growth of this sustainable industry. Embrace the versatility of hemp and join the movement toward a healthier, more sustainable future!

The Plant That Will Help Save the World

In "The Plant That Will Help Save the World," we have journeyed through the rich history, diverse benefits, and future potential of hemp. This versatile plant has been utilized for thousands of years, yet its value is only beginning to be fully recognized in the context of modern challenges. As we confront unprecedented global issues—ranging from climate change and environmental degradation to economic instability and public health crises—it's time to look to hemp for innovative solutions.

Hemp offers a myriad of applications that can address these pressing challenges. Its ability to sequester carbon makes it a powerful ally in our fight against climate change, while its use in sustainable textiles, biodegradable materials, and eco-friendly construction can help reduce our reliance on harmful, non-renewable resources. Additionally, the nutritional and medicinal benefits of hemp seeds and CBD products provide opportunities for improving health and well-being in communities around the world.

However, the journey toward fully realizing the potential of hemp requires collective action. We must advocate for policy reforms that facilitate hemp cultivation, educate ourselves and others about its benefits, and support the growth of a robust hemp industry. By fostering a culture of acceptance and innovation around hemp, we can dismantle the outdated stigmas that have long hindered its progress.

The path forward is clear: hemp can indeed help save the world—if we let it. By embracing this ancient plant and integrating it into our agricultural, economic, and social systems, we can take significant steps toward a more sustainable and equitable future. Let us be inspired by the potential of hemp and work together to unlock its myriad possibilities for the benefit of our planet and future generations.

As we conclude this exploration, let us carry the message of hemp's promise into our communities, advocate for its acceptance, and participate in the movement toward a more sustainable world. Together, we can harness the power of this remarkable plant and pave the way for a brighter, greener future.

Sending a letter to your government officials accompanied by a copy of this book can be a powerful catalyst for change for several reasons:

1. Elevating Awareness

By providing a well-researched and comprehensive resource like this book, you are equipping government officials with valuable information about the benefits and potential of hemp. Many policymakers may not be fully aware of the various applications of hemp or the economic, environmental, and social advantages it offers. This book serves as an educational tool that can broaden their understanding and foster informed discussions about hemp policy.

2. Personal Advocacy

A personalized letter demonstrates your commitment to the cause and illustrates that there are constituents who care deeply about hemp and its potential. Personal stories or insights included in your letter can resonate with officials, making the issue more relatable and urgent. When they see that their constituents are engaged and informed, they may be more motivated to take action.

3. Encouraging Policy Reform

Policymakers respond to the needs and concerns of their constituents. By sending a letter that outlines the necessity for policy reforms to support hemp cultivation and utilization, you are directly advocating for change. Providing a copy of the book can help substantiate your arguments, offering concrete examples, data, and case studies that highlight the success of hemp in various regions and industries.

4. Building Momentum

When individuals take action by writing letters and sharing resources, it can inspire others to do the same. Your proactive approach can create a ripple effect within your community, encouraging more people to advocate for hemp and engage with their own representatives. This collective effort can amplify the demand for change and create a stronger, unified voice in support of hemp policy reform.

5. Demonstrating Public Support

Sending letters to government officials signifies public interest and support for hemp initiatives. When officials recognize that there is a demand for hemp-related policies and a growing constituency advocating for sustainable practices, they may be more inclined to prioritize hemp legislation in their agendas. Public support can be a powerful motivator for policymakers to take action.

6. Fostering Collaboration

By sharing this book, you may also open the door for dialogue and collaboration between government officials and hemp advocates, farmers, and businesses. This can lead to partnerships that promote research, innovation, and the development of a robust hemp industry. Encouraging officials to participate in discussions, conferences, or community events focused on hemp can further enhance their understanding and commitment to the cause.

7. Sending a letter to your government officials

Sending a letter to your government officials along with a copy of this book can make a profound change by elevating awareness, encouraging informed discussions, advocating for policy reform, building momentum for collective action, demonstrating public support, and fostering collaboration. By taking this initiative, you not only contribute to the cause of hemp advocacy but also empower your community to embrace sustainable and innovative solutions for a better future. Together, we can harness the potential of hemp to make meaningful progress toward addressing the challenges we face today.

Here is a sample letter you can use or create your own.

[Your Name]

[Your Address]

[City, State, Zip Code]

[Email Address]

[Date]

[Recipient's Name]

[Recipient's Title]

[Office or Organization Name]

[Office Address]

[City, State, Zip Code]

Dear [Recipient's Name],

I hope this letter finds you well. I am writing to you as a concerned citizen who believes in the potential of hemp to address some of the most pressing challenges we face today. As we navigate a world grappling with climate change, economic instability, and public health crises, it is imperative that we explore sustainable solutions that can drive positive change.

Hemp, an ancient plant with a rich history, offers a multitude of benefits that can play a crucial role in shaping a more sustainable future. Its remarkable versatility allows it to be used in various applications, including sustainable textiles, biodegradable materials, health supplements, and eco-friendly construction. Beyond these uses, hemp also has the unique ability to sequester carbon, making it a powerful ally in our fight against climate change.

However, the full potential of hemp cannot be realized without comprehensive policy reform and public support. I urge you to advocate for legislation that promotes hemp cultivation and removes the outdated barriers that have historically hindered its growth. This includes simplifying regulations, establishing clear guidelines for farmers, and increasing funding for research and development in hemp-related industries.

Furthermore, fostering public awareness and education about the benefits of hemp is essential. By equipping communities with knowledge and resources, we can shift perceptions and encourage the adoption of hemp-based practices in everyday life. This collective effort can help create a vibrant hemp industry that not only generates jobs and stimulates local economies but also contributes to environmental sustainability.

As world changers, we have the power to shape policies that prioritize the health of our planet and its inhabitants. By embracing hemp and its myriad applications, we can take significant strides toward a more sustainable and equitable future for all. I encourage you to consider the potential of hemp as a vital component of our strategy for addressing the challenges we face.

Thank you for your attention to this important matter. I look forward to seeing positive changes that support the growth of the hemp industry and promote sustainable practices. Together, we can harness the power of this remarkable plant to create a better world for generations to come.

Sincerely,

[Your Name]

[Your Title/Organization, if applicable]

[Your Contact Information]